Project M²: Mentoring Young Mathematicians

Sizing Up the Lily Pad Space Station

Measuring with the Frogonauts

M. Katherine Gavin, *University of Connecticut*
Tutita M. Casa, *University of Connecticut*
Suzanne H. Chapin, *Boston University*
Linda Jensen Sheffield, *Northern Kentucky University*

Student Mathematician's Journal

Connections to the Common Core State Standards

Kendall Hunt
publishing company

Credits:
Goliath frog on pages 11, 17 Paul Zahl/Science Photo Library; Bullfrog on pages 13, 27, 29 © 2012 Gerald A. DeBoer. Under license from Shutterstock, Inc.; Tree frog on pages 27, 29 © 2012 Anneka. Under license from Shutterstock, Inc.; Wood frog on pages 27, 29 © 2012 Gerals A. DeBoer. Under license from Shutterstock, Inc.

This material is based upon work supported by the National Science Foundation under Grant No. 0733189 and developed at the University of Connecticut. Any opinions, findings, and conclusions or recommendations expressed in this material are those of the author(s) and do not necessarily reflect the views of the National Science Foundation.

University of Connecticut
Neag School of Education

Kendall Hunt
publishing company

www.kendallhunt.com
Send all inquiries to:
4050 Westmark Drive
Dubuque, IA 52004-1840
1-800-542-6657

ISBN 978-1-5249-1012-9

Published in the United States of America

17 16 15 14 13

Production Date: 2016
Printed by LSI
United States of America
Batch number: 441012

Table of Contents

Table of Contents

As **Speakers** we will:

1. Talk loudly.

2. Turn to the class.

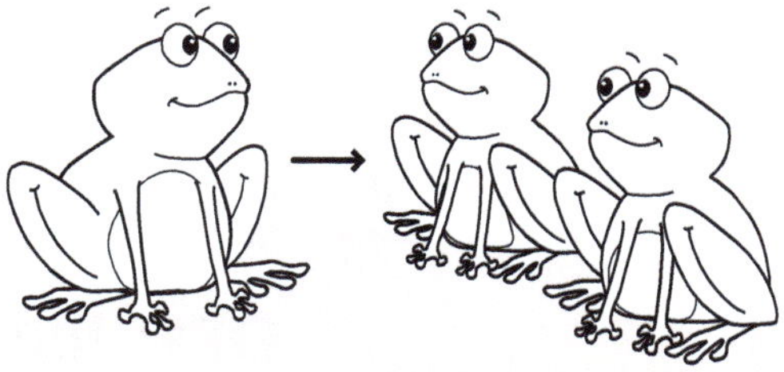

3. Share and explain our ideas.

4. Agree and disagree with ideas, not with each other.

As **Listeners** we will:

1. Ask speakers to speak up.

2. Show speakers we are listening.

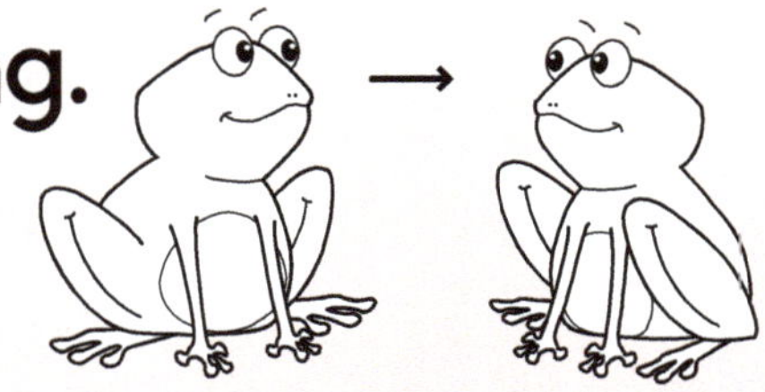

3. Ask questions to understand an idea.

Did you mean...?

As Writers we will:

Think about the question.

Talk about the answer.

Tell

1. all ideas
2. the answer
3. "why"

Student Mathematician: ______________________________

THINK DEEPLY

Pick a tool.

Draw a picture to show me how you used it to measure.

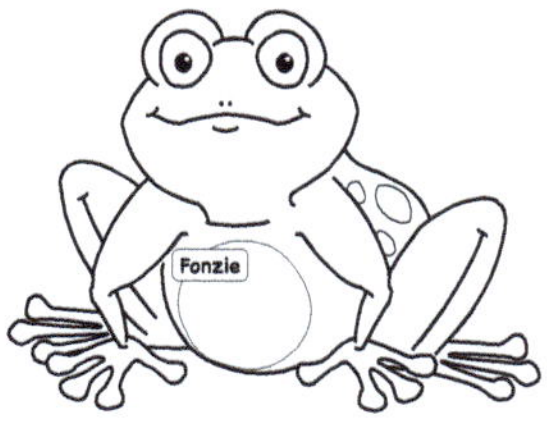

Tell me about your picture.

____

____

Student Mathematician: ______________________________

Shortest to Longest

Space Lab

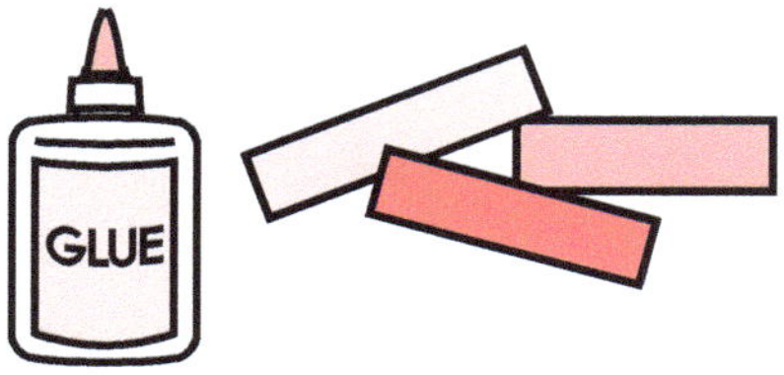

Glue the strips from **shortest** to **longest**.

shortest

longest

Write an idea about the strips using the word **shortest** or **longest**. ______________________________

____

____

Student Mathematician: ____________________

Fonzie thinks the black dot strip is <u>shorter</u>.

Do you agree? Yes No

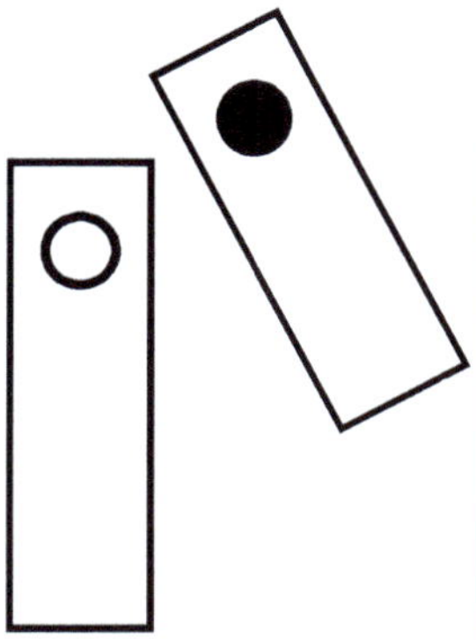

Draw and write why you agree or disagree.

Student Mathematician: ____________________

Shuttle Flight Sheet

Whole-Class Investigation

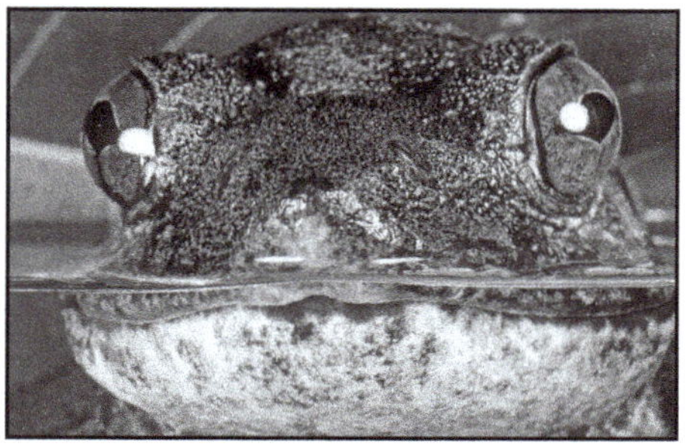

Goliath Frog

○ Green

Predict ________ goliath sticks

Measure ________ goliath sticks

○ Red

Predict ________ goliath sticks

Measure ________ goliath sticks

Student Mathematician: ______________________________

Shuttle Flight

Space Lab

Bullfrog

Predict ________ bullfrog sticks

◯ Green

Measure ________ bullfrog sticks

Predict ________ bullfrog sticks

◯ Red

Measure ________ bullfrog sticks

Student Mathematician: ______________________________

Shuttle Flight

Space Lab

○ Blue

Predict ________ bullfrog sticks

Measure ________ bullfrog sticks

○ Yellow

Predict ________ bullfrog sticks

Measure ________ bullfrog sticks

Which tape is **longest**? ______________________________

Which tape is **shortest**? ______________________________

Student Mathematician: ______________________________

Shuttle Flight Sheet

Whole-Class Debriefing

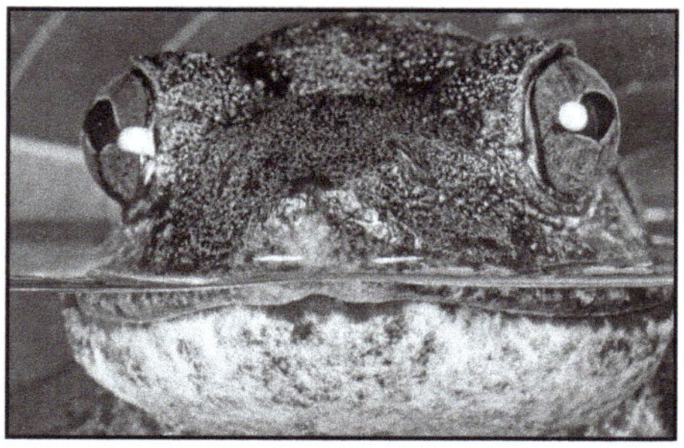

Goliath Frog

○ Blue

Predict ________ goliath sticks

Measure ________ goliath sticks

○ Yellow

Predict ________ goliath sticks

Measure ________ goliath sticks

Student Mathematician: ______________________________

THINK DEEPLY

I am confused. I measured and got different answers! This is what I did.

A.

B.
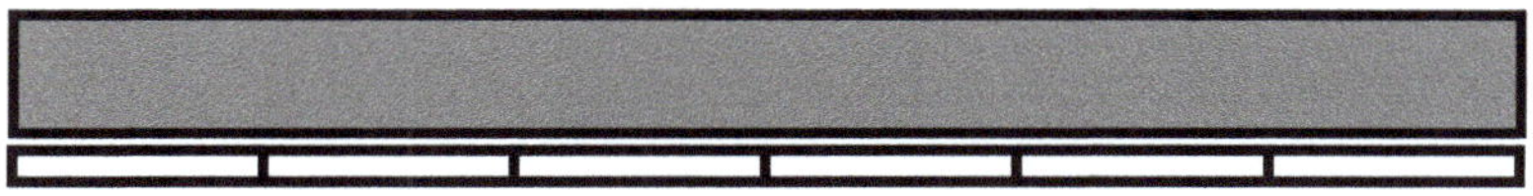

C.
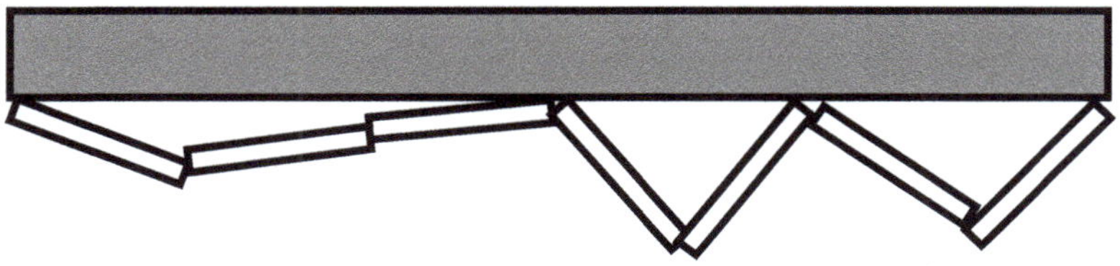

D.
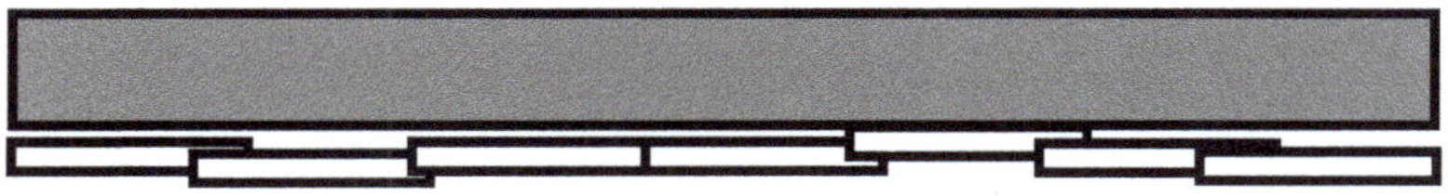

1. Circle the right way to measure.
2. Explain why two of the others are wrong.

_____ is wrong because ______________________________

____.

_____ is wrong because ______________________________

____.

Student Mathematician: ____________________

How Many Feet Can a Bullfrog Stretch?

Space Lab

1. How many  ? ____________________

2. How many 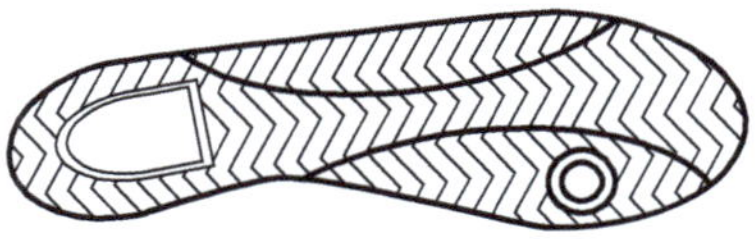 ? ____________________

3. Circle.

I used **more**   .

Freeda

Student Mathematician: ______________________________

THINK DEEPLY

1. Circle the strip that used the <u>most</u> feet to measure.

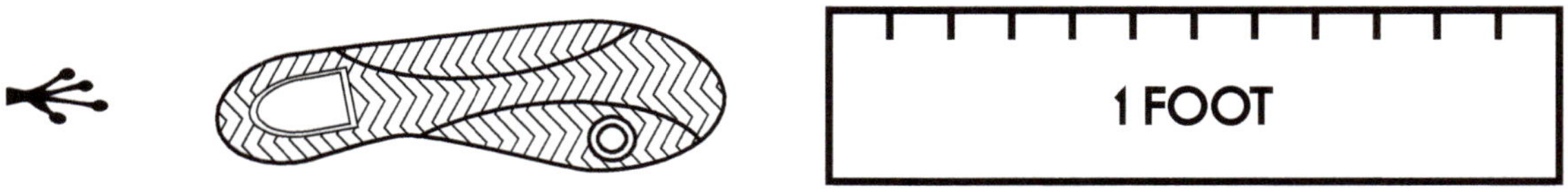

2. Circle the measuring tool that is the <u>shortest</u>.

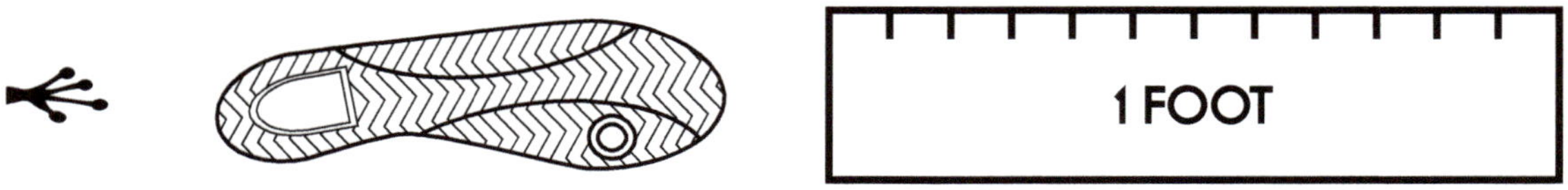

3. Circle the strip that used the <u>least</u> number of feet to measure.

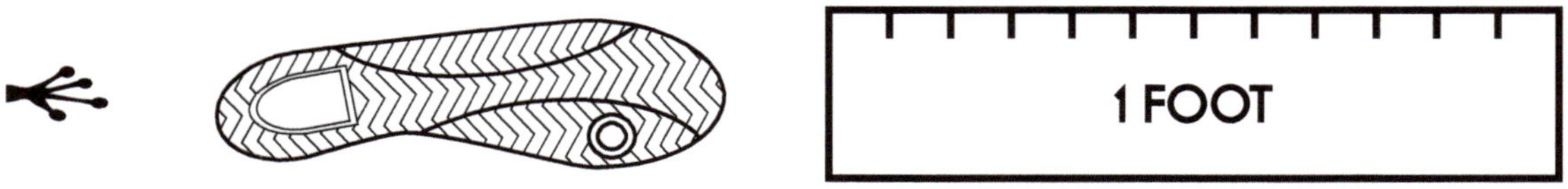

4. Circle the measuring tool that is the <u>longest</u>.

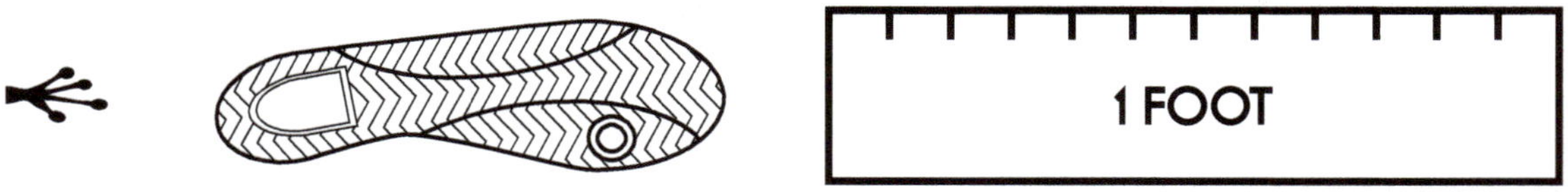

Student Mathematician: ______________________________

THINK DEEPLY

Draw and tell Freeda why the measurements are different with different units.

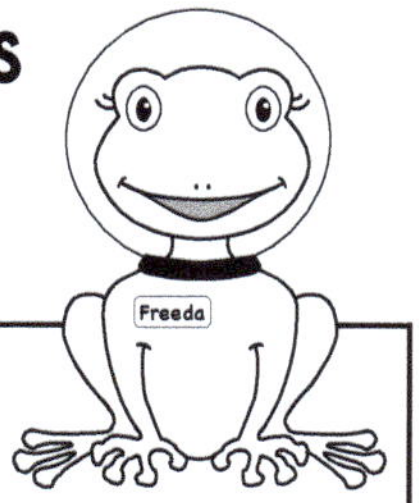

Dear Freeda,

____

____

____

____

Your friend,

____

Student Mathematician: ______________________________

How Many Lily Pads?

Space Lab

1. How many  ? ______________

bullfrogs

2. How many  ? ______________

tree frogs

3. Predict.

There will be **more** / **less**  than  .

wood frogs — tree frogs

Student Mathematician: ______________________________

THINK DEEPLY

I wonder which frogs would have the most lily pads covering the resting pod floor.

Circle one.

bullfrogs

tree frogs

wood frogs

Tell me why!

____

____

____

____

Student Mathematician: ______________________________

Freeze-Dried Flies

Space Lab

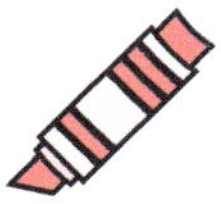

Color in each ○. Measure the volume of each box.

Predict.

______ cubes

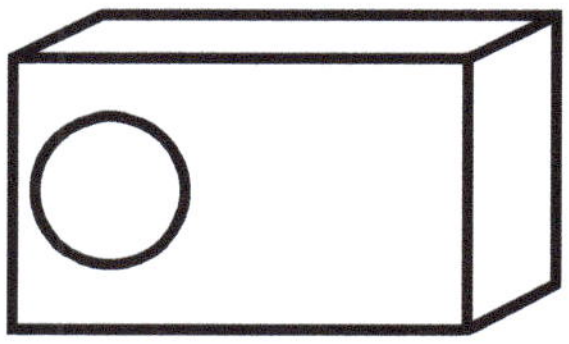

The volume is ______ cubes.

Predict.

______ cubes

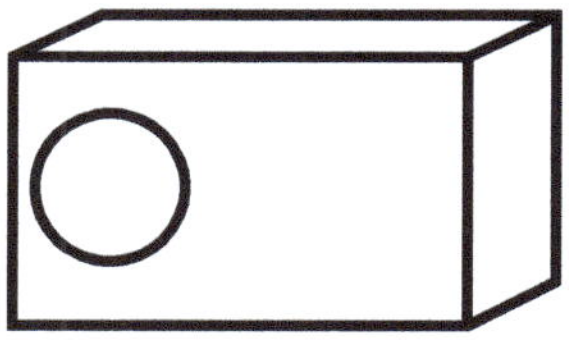

The volume is ______ cubes.

Use the words **more** or **less**.

The ○ box holds ______________________.

The ○ box holds ______________________.

Student Mathematician: ______________________________

THINK DEEPLY

I am confused.

When I measure I get different answers.

What am I doing wrong?

1. ______________________________

____

____

____

2. ______________________________

____

____

____

Student Mathematician: ______________________________

Measuring Capacity

Space Lab

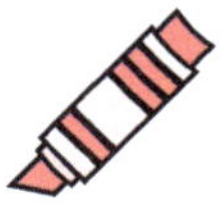

Pour to measure. Color in each ◯.

1. The 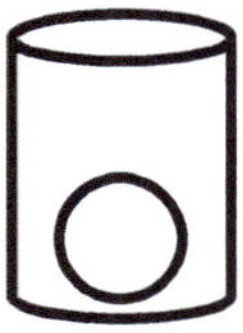 container holds __________ 1-cup scoops.

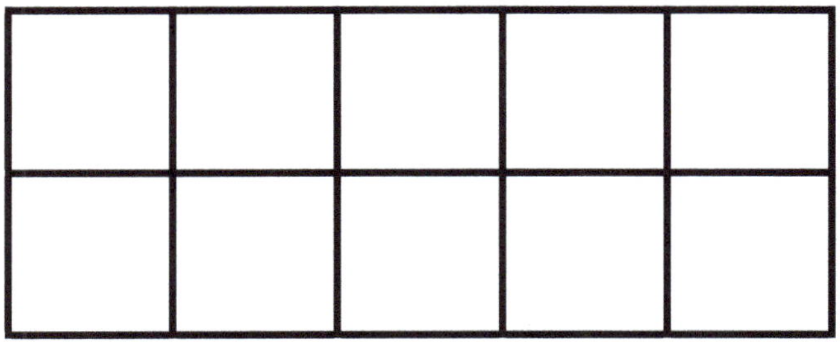

2. The 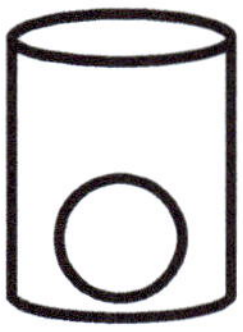 container holds __________ 1-cup scoops.

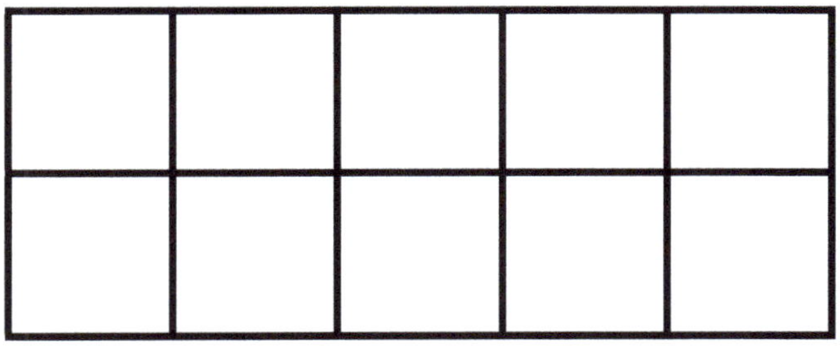

3. The  container holds **more**.

4. The  container holds **less**.

Student Mathematician: ____________________

Which Holds More?

Pour the same amount into each container.

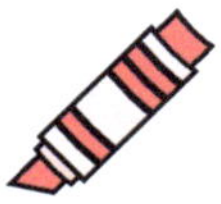

Color how full the container is.

________ 1-cup scoops ________ 1-cup scoops

The 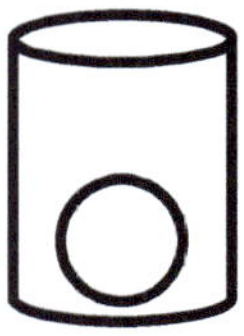 container holds **more**.

The 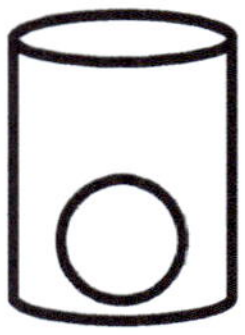 container holds **less**.

Student Mathematician: ______________________________

THINK DEEPLY

I am confused! Which one holds more?

Draw the containers.

Show how much water is in each one.

______ 1-cup scoops	______ 1-cup scoops

Circle the one that can hold more water.

How do you know?

____

____

____

Student Mathematician's Glossary

______________	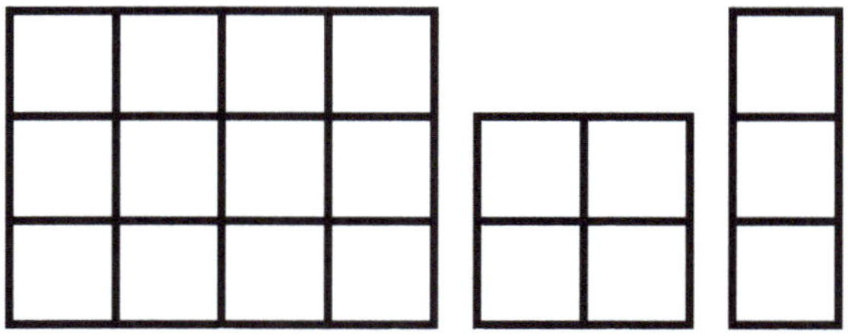
______________	
______________	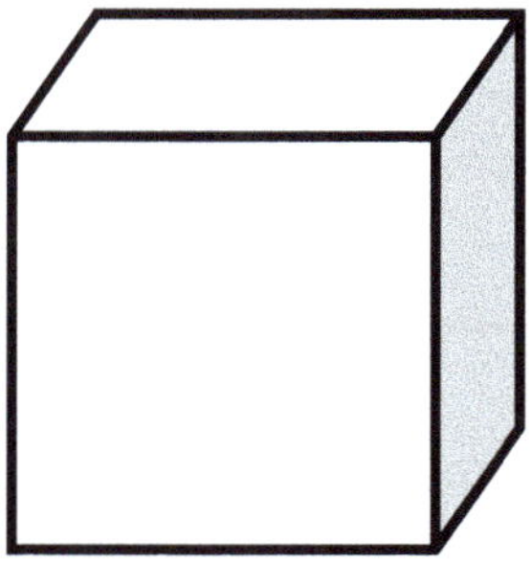
______________	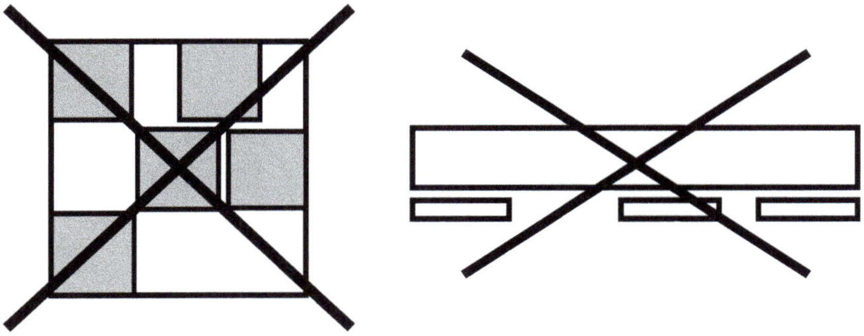
______________	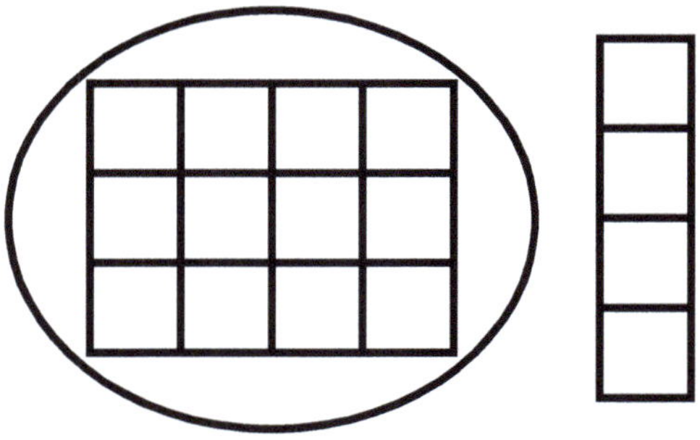

Student Mathematician's Glossary

	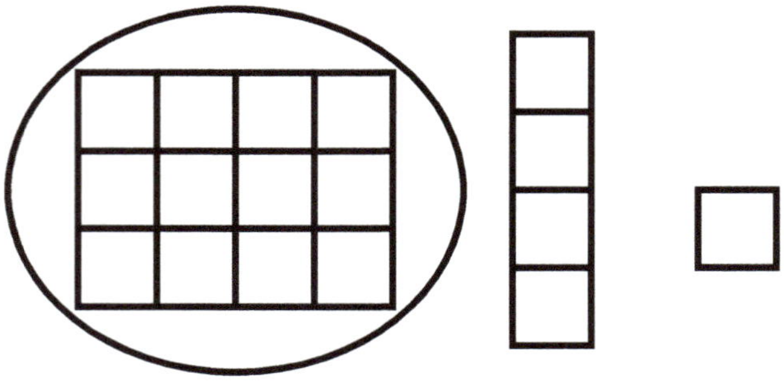
	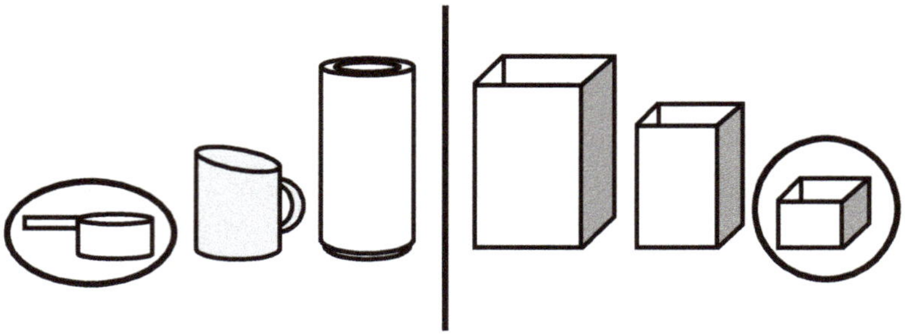
	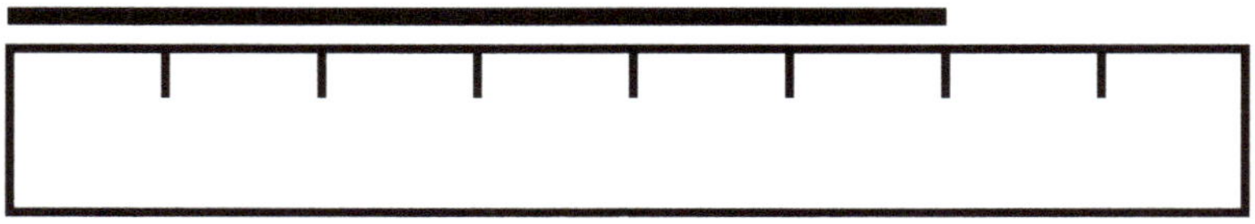
	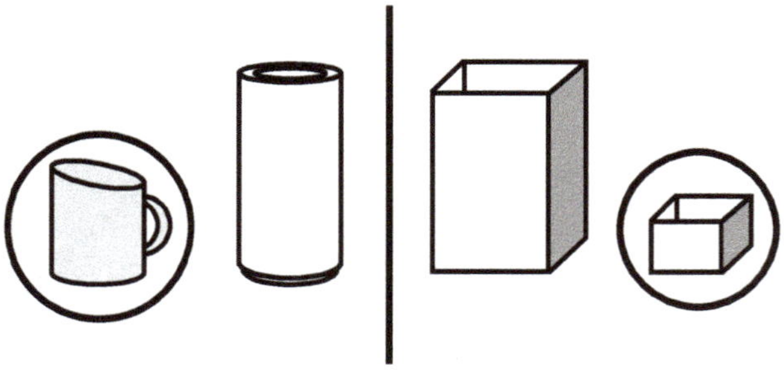
	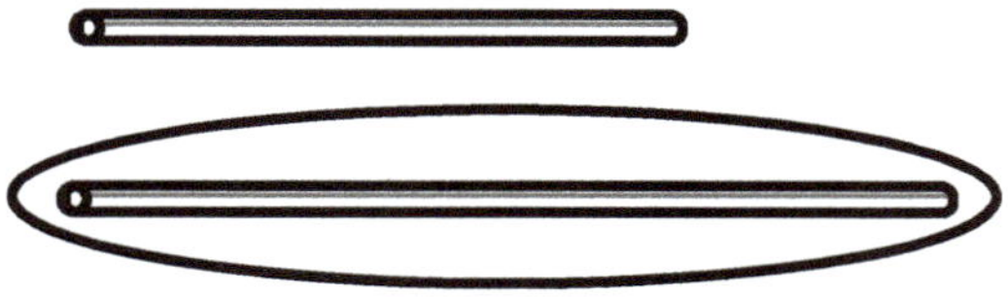

Student Mathematician's Glossary

______________________	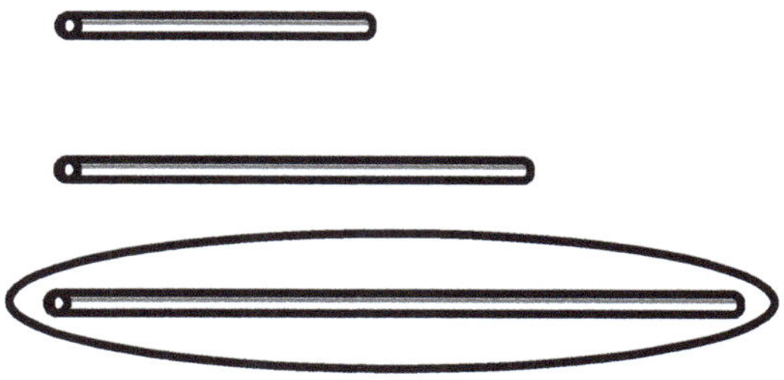
______________________	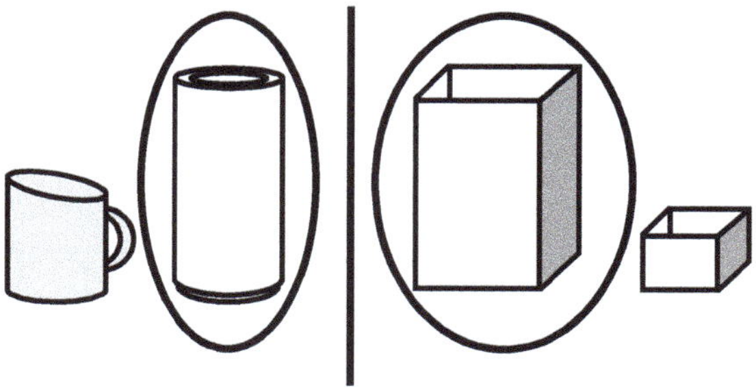
______________________	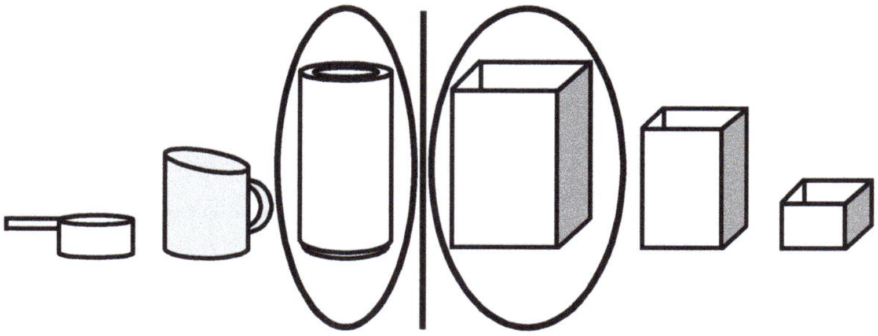
______________________	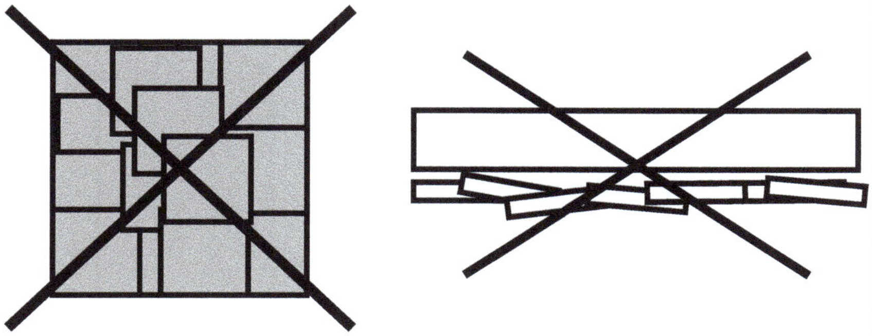
______________________	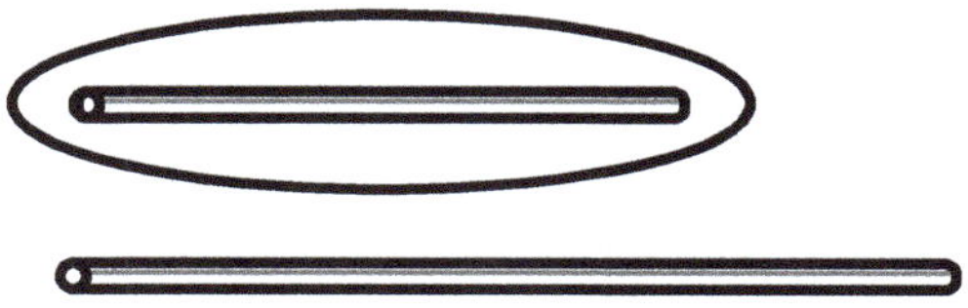

Student Mathematician's Glossary

______________	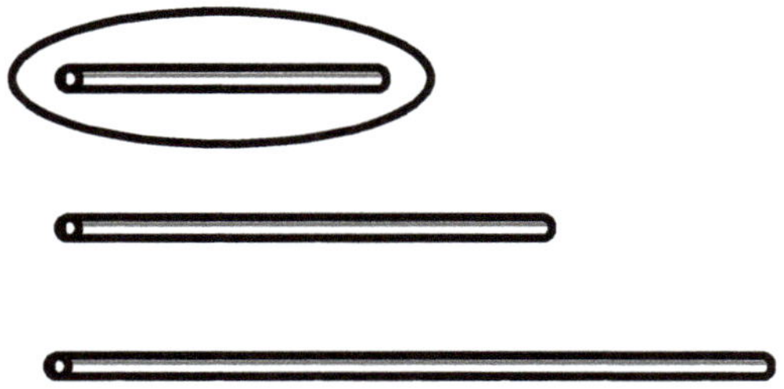
______________	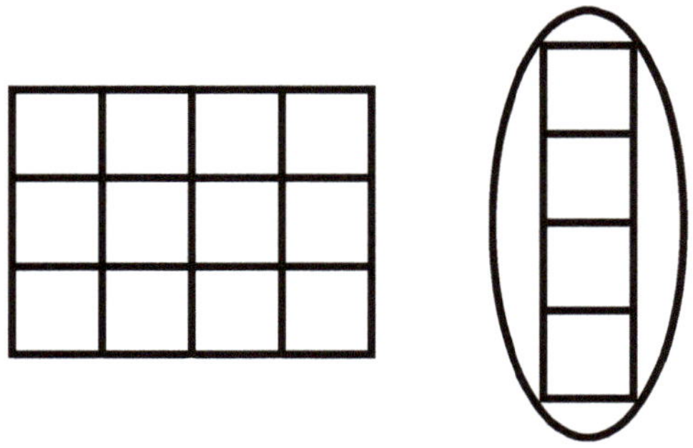
______________	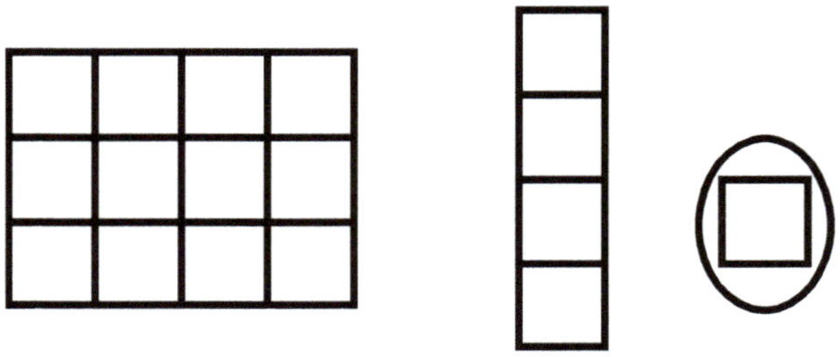
______________	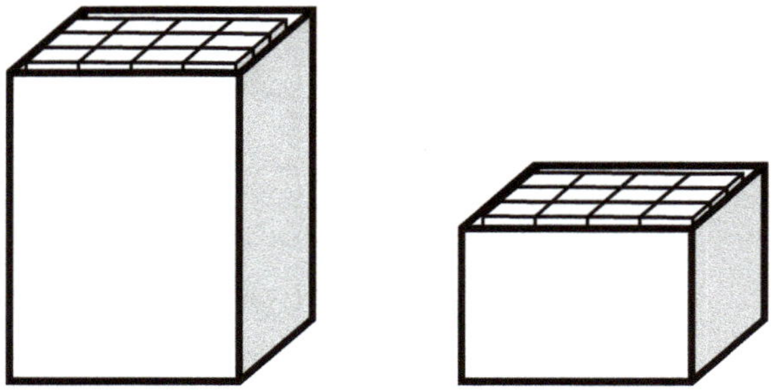

www.ingramcontent.com/pod-product-compliance
Ingram Content Group UK Ltd.
Pitfield, Milton Keynes, MK11 3LW, UK
UKHW050722170926
15229UKWH00010B/47

9 781524 910129